Happy Chris

Love from

Audrey & Stan

PICTURES FROM BARK

Designed, typeset and composed in Australia
Printed by the Everbest Printing Co. Ltd., Hong Kong

PICTURES FROM BARK

by

NANCY MILLARD

CRESCENT BOOKS
NEW YORK

Published by The K. G. Murray Publishing Company Pty. Ltd.,
142 Clarence Street, Sydney, N.S.W.

1st Edition, 1970
1st Revised Edition, November, 1972
2nd Reprint, July, 1974

SBN 85566 003 1

By the same author . . .
"DRIED FLOWER ARRANGEMENTS"
"HAVE FUN WITH DRIED FLOWERS"

Contents

ACKNOWLEDGEMENTS

While I was working on this book my family called our large sun room "The morning after Cyclone Anna". With seven cartons of paper bark scattered over the floor and furniture, together with all the many additions used in making bark pictures, this name has described it well.

During the preparation of this book my late husband, Gilbert, and my sons, Richard and Philip were helpful in every way. They gave me encouragement and constructive criticism, also mechanical help when it was required. Added to this they patiently put up with the inconvenience of a bark-dominated house.

As with my previous book, "Dried Flower Arrangements", my husband carefully photographed each article as I constructed it, as well as the various bark pictures and baskets lent to me for inclusion in this book. For his photography and other help I am deeply grateful.

I am grateful also to Mrs Nina Martindale, of Sydney, who introduced me to bark pictures and to those friends who so readily gave me permission to reproduce their work in this book. My sincere thanks go to Mrs Alice Somerville, of Lismore, NSW, who at the age of 95 has a vital interest in all crafts and does top quality work with paperbark and hard plastic.

To all those friends who assisted me in any way in compiling this book, I give my sincere thanks.

Introduction

When the first paint brush is put into a child's hand, a new world is opened up—the control of colors. The results may not produce a masterpiece but the child is, at last expressing its freedom to choose where each color is to be placed and how the picture will finally look.

A few children continue this love of painting and develop it, but it lapses with most of them and they never paint or enjoy painting in later life.

During the last 20 years in almost every town and suburb in Australia there has been a great upsurge of interest in creative painting. Art groups have been formed, sometimes in hamlets so small that a sign post on the road side is the only indication of habitation. Membership of these art groups usually increases rapidly, an art show is arranged and the future of the organisation is assured.

The members of these art groups are mainly women, but many men find that they too have the urge to paint. The relaxation of painting is as important as its creative satisfaction and the gradual visible success in producing a picture of interest and beauty.

On the other hand there are many people who have no wish to paint or who lack the confidence to join these groups, but who have a strong love of beauty and a latent urge (perhaps unknown even to themselves) to make a picture beautiful enough to adorn their homes and give pleasure to those who see it.

It is with this group even more than with those already occupied with painting, that bark picture making could reveal a new sphere of pleasure and achievement.

The effort involved is not very great and the prospective picture maker's own assessment of her lack of "artistic talent" is not always of importance. So often the talent is there, but it is dormant, not dead. The really surprising thing is how this talent can be stimulated and how it thrives and blossoms when it is put to a test.

As the simple techniques are mastered, the time and effort decreases with each picture made. Children are particularly good at bark pictures and frequently need little or no help in producing a surprisingly attractive one. They like the idea of building a picture from already formed and colored hills and cliffs and have definite and original ideas of the effect they are striving for. They sit absorbed and quiet with only occasional demands for approval or help which is natural in any constructive occupation.

There is a young quadraplegic girl living in Tasmania with whom I have corresponded about the making of dried flower pictures. She is interested in all craft work and is able to produce quite incredible things with the aid of an instrument held between her teeth. With no assistance other than my written directions and a good assortment of bark, she was able to make bark pictures. Her mother helped her only in arranging the material in an accessible position.

Hearing of this young girl's success under such crushing difficulties may give courage to a timid starter. So many women have told me that they lack all artistic talent and could never make a picture. The hardest part is to get them actually to try. The rest is easy and with a lot of encouragement and a little help they invariably go on to produce a picture which not only astounds the maker but will delight her family when it is shown to them.

If you live in an area where paperbark trees grow and have a few old frames stored somewhere in the house, then the expense of making bark pictures will be nil. For those who live long distances from the coast, bark is sold cheaply by many charitable organisations. Usually one carton costing about two dollars will be enough for 10 or 12 people to make their first pictures.

The most widely used bark comes from the paperbark tree called the melaleuca; bark from the leptospermum can be used too, though their covering is more meagre and harder to gather. Both varieties are to be found growing round the coast of Australia and in some inland areas. They grow well in brackish soil and swampy flats, but seldom reach any great height. Their leaves are narrow and are usually small, but vary in size and shape according to the species to which they belong.

The red browns of the Central Australian background blend well with the Aboriginal hunters in this fine picture made from bark. Made by Zillah Needham, Orange, N.S.W.

The name "teatree" was given to the leptospermum and some of the melaleucas by our early settlers, who, when supply ships were delayed and stores were scarce, dried the leaves of these trees and used them for making tea. Nowadays branches and leaves of leptospermum varieties are used to thatch the roofs of bush houses; the small branches make attractive brush fences for gardens and a rough type of witch's broom.

The bark of the melaleuca is made up of layer upon layer of thin papery sheets reaching at times several inches in thickness. Frequently there is an astounding variation in color in the various layers.

The bark varies greatly in color according to the locality in which it is growing. Highland trees produce quite a different range of colors from the swamp grown melaleucas. Climate and season also have their influence; long periods of wet weather on a piece of bark which has fallen on cool damp earth will produce green shades, from palest apple to really vivid greens. This color is formed by lichen growth; lichen

9

also grows on old bark which has loosened and will soon fall from the tree.

Paperbark trees shed their covering about twice a year, but not always at regular intervals. Peeling this loose bark from the tree will not hurt it, but cutting with an axe can cause a severe check and perhaps even the death of the tree.

I have frequently seen trees cut deeply with an axe and the bark peeled to the hard wood. When the sapwood is exposed in this way to our extreme weather conditions, the tree must receive a severe shock. A bushfire coming at this time when the tree's natural covering has been removed would surely kill it. The many layers of thin papery bark are a superb insulator and burning rarely penetrates them deeply. So melaleucas have an excellent natural protection against summer heat and the ravages of fire.

When a bushfire rages through Australian bushland it sometimes travels so quickly that the trunks of the trees are only badly singed while all the tree's foliage and most of the branches are burned to ashes. Fortunately, these trees frequently recover and sprout new shoots after the first shower of rain. When melaleucas are lightly burned in this way, the resulting blackness of their outer covering is useful in a picture. It should be removed from the tree with care as it is brittle and may break easily.

At all times the external color of the bark gives absolutely no indication of the colors hidden underneath. Sometimes the shades of each layer are similar, but it is more usual to find a great variation in one piece of bark. It may be streaked or even spotted or have a gradual deepening of color towards one side.

The color range is wide, starting with clear pure white, through creams, yellows and buffs and all the shades of brown. There are many red tones, too, from pale shell pink to rich salmons and vivid reds, and, as mentioned before, the lichen-formed greens plus the wonderful blacks caused by bushfires. Fuchsia shades are rarer, but turn up now and then, also magentas and purples. These last three are the only colors hard to use. On occasions they give good dramatic contrast, but more frequently they look out of place and clash with the other bark colors.

Some barks are suede-like in texture and are of delicate colors— usually pink and cream—and they enhance any picture. By far the most common shades are buffs and browns and these are usually the pieces left over after the first big picture-making effort.

There are other types of bark that can be combined with paperbark with effect in picture-making. The geebung is one of these. The species most useful is *Persoonia pinifolia,* a tall slim tree, quite often spindly and inconspicuous. It has long thin leaves, yellow flowers and fleshy fruit, edible, but at times astringent. Its bark is dark externally, but a brilliant ruby red inside. It is flaky, rather like puffed pastry, and can be separated into layers with a knife. The bark is easy to remove from the tree, but, as with paperbark, only the outer layers should be taken leaving sufficient fire and heat protection for this slender tree.

In inland areas of Australia there are many other strange and beautiful barks. The leopard tree is attractively spotted, but so often these lovely barks are thick and hard and of little use in a picture.

Permission should always be obtained from the owner before removing paperbark from trees growing on private property. This is usually readily granted if assurances are given that the tree will not be injured. In parks, the ranger may be only too pleased to have fallen bark carried away, but again—ask permission first.

A quick look through a bundle of bark will reveal pieces which immediately suggest hills and mountain peaks, water and stones. A little experiment in putting these ready-made features together is the next step and the one so many people find so hard to take. Nothing is lost if the first attempt is a failure as the bark will be quite unspoiled. Even a very experienced bark picture-maker will find it necessary to try several likely pieces before the perfect one is found.

My advice to you is to have courage and to attempt to make your initial bark picture. Then a few hours later you will be standing back admiring your first beautiful work of art and planning that a minimum of time elapses before you begin your second one.

CHAPTER ONE

Choosing and finding the bark to use

PAPER BARK [OR TEA] TREES of Australia (Melaleucas and Leptospermums) vary considerably in their growth habits according to their environment and species. Once they are identified it seems easy for the enthusiast to pick them out in any assembly of trees.

In the Gulf country of Northern Queensland their growth is rather like that of weeping willows, though more upright. When found along the banks of lakes and rivers, their branches are pendulous and trail in the water in the way that willows do. Away from the water they grow taller and more erect and mature into fine strong trees. In both situations with their deep green shiny leaves, they are very picturesque and beautiful and the layers of their papery bark feel like fine soft suedes.

Further south in the subtropical areas of good rainfall, their growth is more upright and where the locality suits them and the wind is moderate, they can grow into fine specimens.

These trees tolerate sea spray and a saline swampy soil. As this type of soil is usually deficient in plant food and exposed to sea winds, the coastal paper barks in the southern areas of Australia are smallish and sometimes distorted by their environment. The occasional tree growing to any great size has thick layers of bark, but it rarely has the many variations in color found in the smaller, harder living specimens.

The external color of the paperbark tree will give little indication of the shades to be found inside the bark layers. With experience the bark gatherer learns to pick out the more likely trees though this can never be done with certainty. It is easy, though, to find these colors by peeling some bark back or removing a piece or several pieces from the

The pictures on this and the next three pages show the type of trees to look for, with a close-up of their leaves to help you identify them. The tree above is a Melaleuca decora from the Central Coast of N.S.W.

same tree and carefully examining its layers. Then, if useful colors are not present, move on to another tree.

The trees shed unwanted bark several times a year. It is not damaging to the tree to peel more away by hand. Never use an axe as grave injury to the tree could occur by exposing the sapwood to hot temperatures and bushfires. The outside bark of the tree is sometimes fairly rough with its color either bleached white or grey by the sun or darkened by weather or fire.

Fallen bark can be of value, especially for its green shades, but if it has been on the ground for some months and is rotting its usefulness will have deteriorated.

SEPARATING THE LAYERS

Sometimes the layers of a piece of bark will separate easily and will remain almost intact, but other barks seem to have several of their layers fused together or the bark may be coarse and defy separation.

This is a callistemon salignus—the White Bottlebrush—found in Northern N.S.W. and Queensland.

Do not destroy a good piece of bark by trying to reduce its bulk drastically. Four or five layers should not make the picture too bulky to frame, so remove only that bark behind the chosen piece that can be separated without disturbing or destroying its surface.

Occasionally when the inside layers are brick red or white or a mottled combination of these two, the bark will have a powdery covering which loosens with age. If much free powder is present it should be removed by tapping or blowing, otherwise when the picture is being tacked into the frame the powder could break away and adhere to the glass and the picture will be blurred.

The predominant colors in tea tree barks are the pale brown shades and it is this bark that will be left over when the first pictures are completed. Most of it will be well worth keeping to supplement future supplies as the next carton of bark, if coming from a different

A Melaleuca quinquenervia, commonly known as the "broad-leaved paperbark," which is found in many areas of Australia.

locality, may lack a few of the shades and textures present in the first batch.

BEWARE OF TICKS

A carton or sack is a handy container when gathering paperbark and an ardent bark picture-maker will always carry one of these in the boot of her car.

Paperbark is considered to be a possible carrier of cattle ticks and it is illegal to take it from affected areas past the cattle tick controls. A few enquiries beforehand can perhaps save much work and the enforced abandonment of precious bark at the tick control post.

Bark pictures are popular presents for people overseas and as souvenirs. If the completed picture is sprayed with strong insecticide before being sealed in its frame with masking tape, any insect life that may

This is another variety of the widely-spread Melaleuca, which would provide suitable bark. But, please, treat the trees with care!

have survived in the bark layers will be killed. There will then be no danger of introducing undesirable insects into an overseas country.

DRYING AND STORING

If the bark is muddy (as can happen after prolonged rains) it should be washed with a hose and dried in the sun. Shelter from wind is important during this drying period as pieces can blow off in all directions and may be lost among the garden plants.

Drying is also necessary when bark is gathered soon after rain and is damp. Scatter it loosely and the sun will do the rest. Drying could take as little as a day in hot summer to a week or more in winter. If the bark is needed urgently, controlled artificial heat will do it no harm.

The amount of sun needed for drying will not spoil the bark, but direct and fierce sun's rays over a prolonged period could gradually fade the more vivid colors.

The depth of color in damp bark is richer than in dry bark so do not confuse the loss of a little color during drying with fading.

Right: The central cork covering was removed from coasters and replaced by a more colorful bark background for these wall plaques. A hanging cord was glued to the back of the coaster and the small tough flowers and small seeds pasted in place.

Right: In this table top, black and red geebung bark was used. This bark is flaky and brittle and easily crushed. The leaves on the tree were formed with small particles of this bark, the black pieces giving depth to the red. The green lichen was growing on the foreground bark.

RIGHT: *The delicate skeleton tree shapes and the dark red bushes in the foreground are pieces of deep sea, seaweed which came from Esperance, WA. The cliff is a single piece of rough bark, and the bark the crows are made of is thin and soft and blackened by burning. The dead tree's bark is thick.*

BELOW: *Fine black seaweed forms the delicate tree in this placid water scene with a whitish rock seaweed behind reddish foreground outcrop.*

Above: *Green lichen forms the foliage on the trees with a different variety for the leaves of the sedges in the left foreground. S p a g n u m moss and part of a West Australian banksia flower are used as rushes.*

Left: *Geebung bark is used at the base of the white trunked trees whose foliage is made from an olive moss. Native grass nodes make* **realistic bulrushes.**

LEFT: *A light-colored bark was pasted to the cardboard backing and the flowers arranged on it. A spot of adhesive is needed behind each flower head as the surface of the bark is smooth.*

BELOW: *The roses were made and together with their twig stems were glued in place on a suitable background. The soft, silky texture and orange-buff shades of the seed pods of the Golden Rain Tree blend well and make realistic rose leaves.*

ABOVE: *This landscape was inspired by the lovely sunset colors in the hills near Queenstown, Tasmania. Years ago mining fumes killed the trees and the bare hills turn many different colors at sunset.*

RIGHT: *The tinting with poster colors of the wing feathers and beaks of these birds is an illustration of the restrained use of additional color. The bark used for the background and the bodies of the birds is in its natural color.* (Made by Marion Wilson, Oberon, NSW.)

Sixty different colors and color combinations of paper bark are shown above. The pictures on this page are to show you the many varieties of bark and materials that can be used.

Top Right: Mosses and lichens have fascinating shapes and a wide variety of colors useful for tree foliage or ground growth.

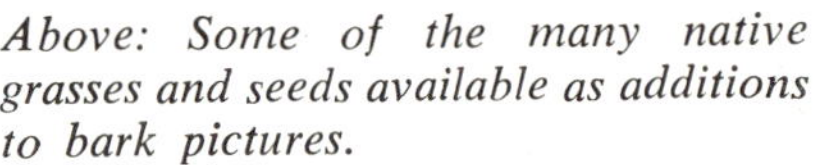

Above: Some of the many native grasses and seeds available as additions to bark pictures.

Left: Seaweeds and sponges in their many shapes and colors may be used to decorate bark pictures.

CHAPTER TWO

The things you need to make a start

Now THAT YOU HAVE made a collection of bark in an assortment of colors and textures, plus various lichens, mosses and so on (see Chapters One and Five), you are ready to make a start on your picture. Here are the materials you will need (as shown in the picture below):

(1) Small brush—a camel hair mop, small paint or pastry brush.
(2) Sharp scissors.
(3) Matches, candle in fireproof holder and damp cloth.
(4) Pair of forceps.
(5) Paste or gum and saucer.
(6) Sheet of firm paper or light cardboard.
(7) Frame and glass, tacks, small hammer and masking tape.

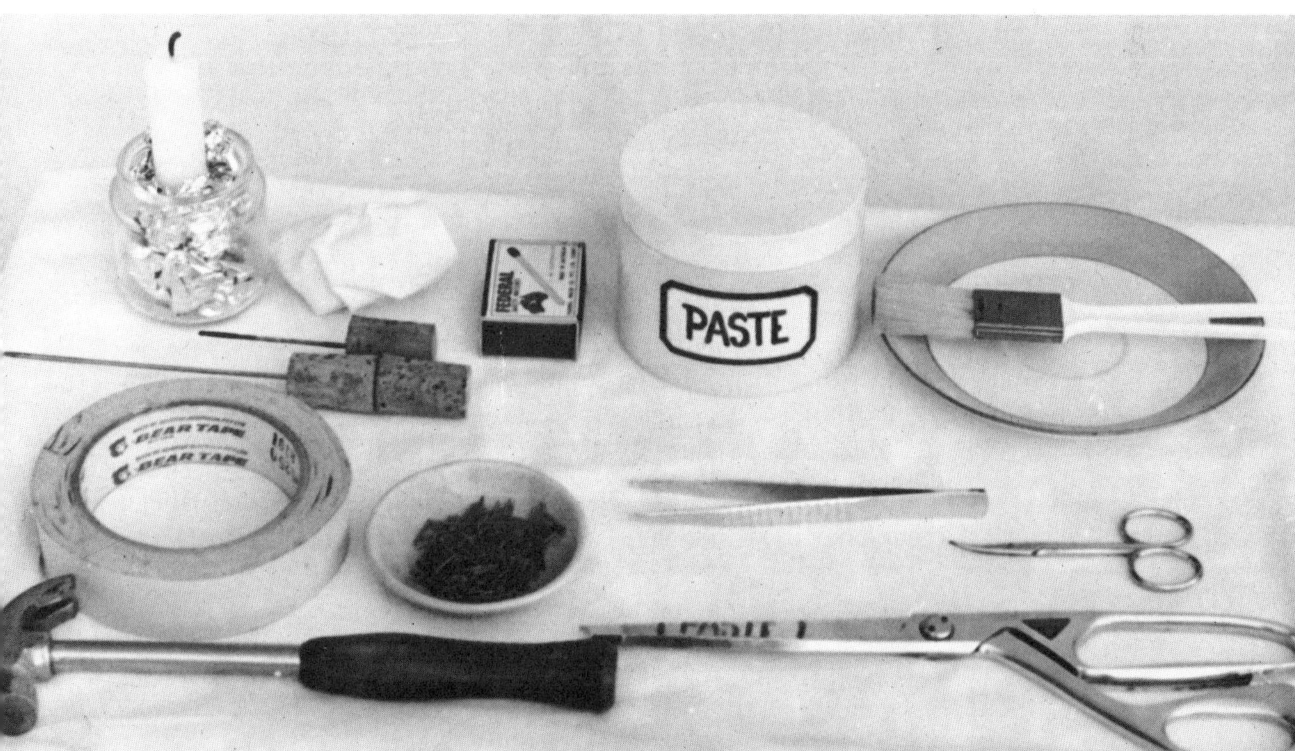

This bark was cut to show the nine different layers which form the tree's thick covering. On some trees, the layers are few and adhere tightly to the trunk.

To make paste

Any thin commercial type of paste or gum can be used, but home-made flour paste is inexpensive and quite satisfactory.

Mix one level tablespoonful of plain flour with two cups of cold water and bring to boiling point, stirring constantly, simmer for three or four minutes. Remove saucepan from flame and stir in one teaspoonful of kerosene if silverfish are troublesome in the district. Strain if lumpy and pour into a wide, low screw-top jar. If the paste is to be used again store in a refrigerator. It should keep for two to four weeks depending on the season.

Lighting is important

A good light is necessary for bark picture-making and daylight is a little easier to work in than artificial light.

A kitchen with a strong light is a suitable hobby room, but if the dining room is more convenient, then use it; cleaning up when the picture is completed is no problem. A sheet or newspaper will protect the table and a vacuum cleaner will soon pick up all the dust and any small pieces of bark on the carpet.

Face the light or have it coming in from either side rather than from behind. Back lighting will form a body shadow over the work space and the subsequent need to move the picture into direct light to assess the value of each additional piece of bark can be frustrating.

For a final and critical examination of the finished picture an even better light is needed. Sunlight or close-up electricity will quickly reveal small faults.

THE PAPER TO USE

Colored paper for the picture's foundation is unsuitable as small holes normally unnoticeable in the bark could show an alien color. White paper is the most satisfactory and does not dull the brightness or change the color of the more flimsy pieces of bark. Firm drawing paper is ideal.

Cut the paper the exact size of the glass—precision at this stage will make the final trimming of any overlapping bark much easier.

SETTING UP THE WORK TABLE

It is usually more convenient to spread the bark out on one table and work on another. With more than two workers this arrangement would be a necessity.

I always have several worn out, but not yet "holey" sheets to use when I carry large quantities of flowers in the boot of the car. One of these sheets is ideal to spread over a table before the carton of bark is emptied on to it and another to cover the work table. The big advantage in using a sheet is that my whole working mess can be rolled up and out of sight in a few seconds if the need arises. Most women will admit that this need often does arise especially when the craft worker becomes absorbed in her work and goes well over the time she has allowed herself between household jobs.

After the table is covered with a sheet or newspaper, place the paste pot, the brush and a saucer to hold the brush when not in use towards the centre of the table, with scissors and forceps nearby.

The firm piece of paper cut exactly to the size of the glass should be directly in front of the worker and the frame and glass on her right hand side.

CHAPTER THREE

Selecting or making a suitable frame

NATURAL WOOD BLENDS well with bark colors and plain box frames are readily available from chain stores. Do not choose the "slide-in" type of frame as the extra bulk caused by the bark makes it difficult and at times impossible to force the picture back into the frame. It is possible to remove the overlapping rabbet edge in a slide-in frame with the aid of a chisel or strong sharp knife and then the frame will be quite suitable.

Lift-out frames are much easier to use. Suitable colors vary from natural and the brown shades to black, white, grey, cream and even gold. I feel that new frames give a slightly more finished effect, but if old ones are available they can often be used with excellent results. If shabby, they should be scrubbed and given time to dry before being repainted; the glass should be washed and polished.

An attractive small bark picture can be made in a round white plastic frame (available at chain stores at small cost). The convex glass gives added interest and perspective to the picture. Cut a firm paper circle using the picture's backing as pattern and gum it and the backing together. Construct the picture on this backing using more paste than usual for each piece of bark as the curvature of the glass will give no assistance in holding the bark in place. Add glass and frame and press backing firmly into place.

When bark pictures are to be sent overseas by air and a frame is undesirable because of its weight and the fragility of the glass, the whole picture can be wrapped in soft plastic and sealed at the back with clear adhesive tape. Placed between two pieces of cardboard the picture will then keep rigid in transit.

If there is a handyman or woman about the house, simple frames

*The pictures were made and securely pasted
down as the glass is convex and would not
hold the bark in place. Seaweed forms the
dark tree which stands out effectively.*

are not difficult to construct, but precision and neatness in workman-
ship are essential.

A mitre box, a sharp, true, fine-toothed saw, a few fine nails of the
right length and a hammer are the tools needed, as well as a supply of
picture framing timber and some wood adhesive. Mitre boxes are made
of wood or steel. Their function is to hold the framing wood while an
accurate cut of 45 degrees is made. A wooden mitre box is inexpensive
and quite effective.

Ready-grooved framing timber is available, but lengths of ordinary
builder's quad or scotia can be rabbetted on their reverse side (quad
is convex beading and scotia is concave). This rabbet groove should

The soft plastic covering enables these pictures to be airmailed overseas as souvenirs. They were, made by Aboriginal women at Cowra, ·N.S.W.

be about a quarter of an inch wide and about three sixteenths of an inch deep to hold the picture, glass and backing in place.

Accurately measure each side of the glass and mark these lengths on the reverse side of the wood. The maximum length of the rabbet groove should be equal to the length of the side of the glass. As a precaution when making the first few frames, time should be taken to pencil in the angle of the cut. So often beginners err by accurately measuring and marking the length of the side and then cutting the wood in the wrong direction.

Arrange the framing timber in the mitre box and cut each end at an angle of 45 degrees. Fine nails should now be driven into the corners of the frame until their points are barely showing on the inside of the cut ends. (See illustration). Coat these cut ends with wood adhesive, put them together then drive the nails home.

This operation can be done with the frame held firmly in the vise (well protected from damage from the jaws of the vise) or with the frame on the floor held securely between the maker's knees. Place the finished frame on an even surface. The adhesive will dry overnight and the frame will then be ready for painting or varnishing.

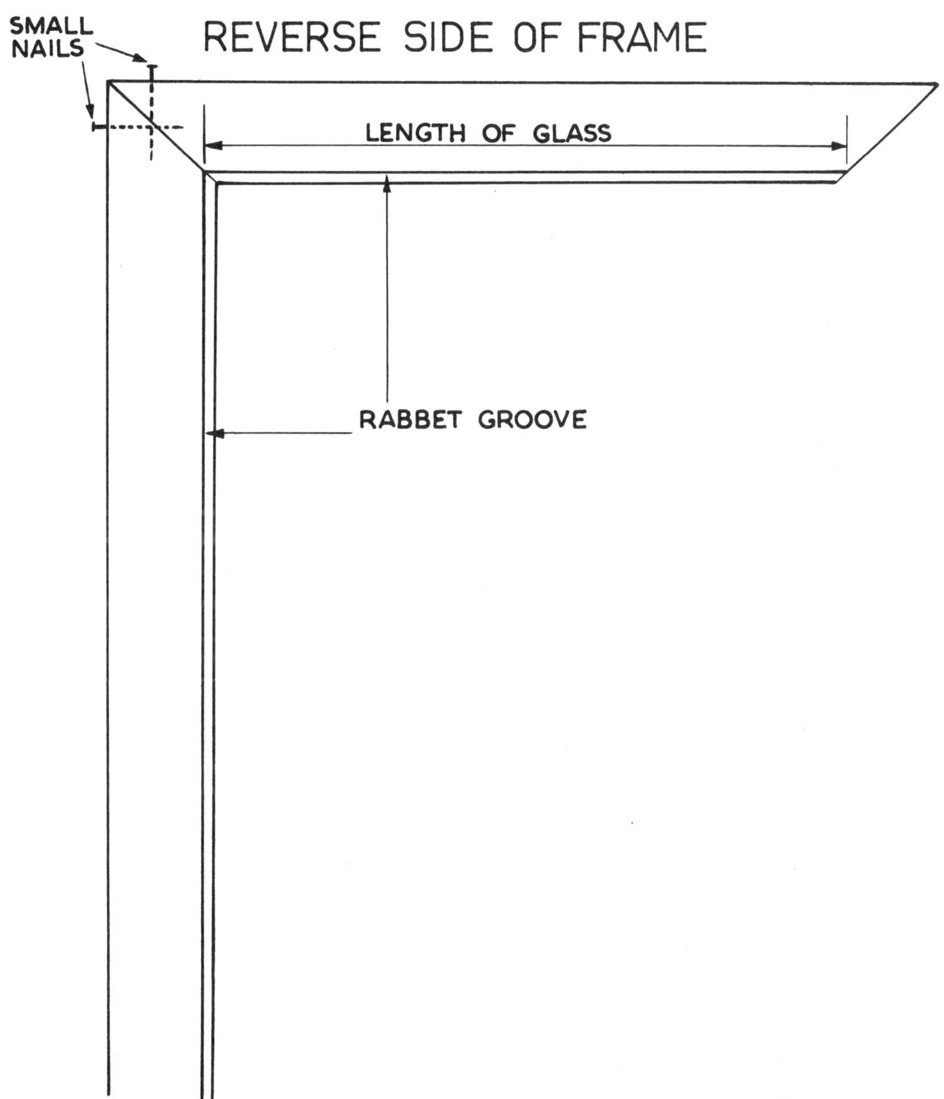

CHAPTER FOUR

Choosing the subject for your picture

A TRUE BARK PICTURE is made with bark in its natural colors and texture and the fine hair-like bark that is sometimes found between its layers. No moss or lichen or materials other than bark should be added.

The second type is made in a similar way to the first but with natural growths, such as lichen, mosses and bush grasses added as decoration. A third type has unrelated natural additions. Seaweed is one of these.

Last of all, for those not altogether happy with the lack of blues and pale shadings, poster colors can be used to tint lightly the sky bark and other areas. Do this with restraint or the picture will look tawdry.

These touched-up pictures could not be entered in any bark picture competition unless a specific clause allows it.

The use of dried pressed delphinium or hydrangea petals for skies would also be excluded.

The need for a blue sky is a personal taste. Having worked so long with the natural colors of bark, I find that when a blue sky turns up unexpectedly in a bark picture, I am reminded of the surprisingly blue eyes of Siamese kittens.

As with painting there is no limit to subject matter with bark pictures, although rural scenes, snow scenes and seascapes are the most popular. Human faces can be made effectively, also animals and large birds. Smaller human figures, horses, sheep, cattle and kangaroos are worth attempting and tree shapes and buildings are not difficult to construct realistically.

Bark is a good medium for abstract pictures.

There are no hard and fast rules for the composition of bark pictures, but a few helpful suggestions for beginners could be of use. These

Small slivers of bark placed side by side form this portrait of a man. The skin color, though pallid, is life-like. Burnt and brown bark form the shadings of the face and hair. Nothing other than bark was used in this picture. Made by Clyde Hicks, Canowindra, N.S.W.

suggestions should be ignored when the maker gains more confidence. They are given only to assist in the construction of the first few pictures and not to cramp and inhibit new ideas and inspirations.

Before starting to make a picture it could be helpful either to draw a rough sketch of the picture in mind or to obtain a suitable landscape print or photograph to copy.

You will soon find that when an attempt is made to copy a picture the final result will only vaguely resemble the original, the main reason for this is that no two pieces of bark are identical. Also when two people try to copy the same picture, invariably the result will be two widely differing creations. On the other hand it can frequently happen that a piece of bark is discovered which is a great improvement on the already planned hill or mountain. When this new piece is used with success, other more interesting barks will be sought and the picture will change its character immediately. It will then be seen that what was meant to be a copy has become a completely new picture and is now an original. Very soon no outside inspiration will be needed.

When deciding the amount of sky area needed for the picture, it is a good guiding rule that most land and seascapes appear to have balance when the skyline is near the upper or lower third of the picture. Later it could be safe to try a centre skyline and achieve success or to have no sky at all. Another suggestion is to keep strong eye-catching landmarks or trees away from the centre of the picture—a little to one side or obliquely across the picture could be more effective. For example, a river or waterfall down the centre of the land mass could cut the picture in two; let them come from either side towards the centre or the lower opposite side.

METHOD OF MAKING SKIES

Two important decisions have to be made before attempting the sky. The first of these is the picture's hanging position. If it is a rectangular frame, is its greatest length to be vertical or horizontal? The second decision is the size of the sky area in relation to the whole picture.

There are several different ways of forming skies when making bark pictures. The first two described here involve the use of the delicate flimsy sheets found between the layers of some types of paper bark.

Large tissue-thin sheets can occasionally be peeled off intact, but more often even the greatest caution will not prevent them from tearing and the pieces may be only one or two inches long. An attractive sky can be made with these small pieces which can now be chosen for their beautiful color variations and transparency rather than size.

Without using any paste, completely cover the sky area of the paper with these flimsy bits, arranging each piece slightly overlapping its neighbour. Keep the color variations as sky-like and as realistic as possible. For example, if it is a sunset sky the bark could be salmon pink where the hills and sky meet, then would gradually fade to a pale pink at the top of the picture.

Forceps are of great assistance when working with fine bark. Fingers can be so clumsy, often displacing more bark than was intended.

Do not be concerned with clouds at this stage; these will be added after the sky is pasted down.

The featherweight pieces of bark must be sheltered from draught —even a beginner's deep sigh will send them floating across the room.

When the sky arrangement is completed and no further improve-

Place a piece of sky bark in position on the paper and press the glass and frame over it to evaluate quickly the final effect.

ment can be made, gently place the glass and frame on top of it. Immediately it will be seen if the desired effect has been achieved.

This arrangement is only a try-out, the next step is to tilt the paper and to slide gently all this bark back on to the table, disturbing it as little as possible. Caution—breathe softly for a few minutes!

Now lightly and evenly paint with paste the whole of the planned sky area on the paper. Too much paste at this stage will ruin a beautifully made sky—will be puggy and will lose all its gossamer lightness.

Replace the bark pieces on the paste-covered paper as near as possible to the first arrangement, patching any uncovered paper with smaller matching bits. These bare spaces are easy to find when the paper is held up to a window or light globe.

Study the sky carefully without the glass and again with the glass in place. Now decide whether the sky is complete or if clouds would be an improvement. Several different shapes of clouds in varying colors should be draped across the sky as a "try on". Guesswork is unnecessary and indeed inadvisable when arranging clouds. It is so easy to spoil an otherwise good sky with too many clouds or with the wrong kind.

It is rarely in a real sky that clouds scatter unditily in different directions; when they do this they are a curiosity rather than picturesque. It can sometimes look effective to have great cauliflower heads of cumulus clouds rising from the horizon or wind-streaked clouds of slightly differing color from the sky.

The wind currents may be more or less parallel to the horizon or coming in from either side in an upward or downward sweep tapering as their formations peter out. It is wise to try them in all these positions before making the final decision as the beauty of the finished sky will have a great influence on the success of the picture as a whole.

There are no difficulties in tearing tapering bark to represent a wind swept sky, as this is the way the frail thin bark will tear away from the larger piece. If a sliver breaks before tapering to nothingness the cut-off look of cloud or sky bark can be avoided by placing any such hard line on the edge of the picture. If this position is unsuitable, then taper or soften this line by tearing or by overlapping the offending edge with a more suitable piece of bark.

One or two light dabs of paste on the sky under the cloud or under the cloud itself will secure it in place. Smooth it into position. There is no need to anchor it firmly as the glass will do this later.

If several large pieces of flimsy bark are to be used, the sky space should be covered by overlapping these pieces—again watch for square ends and eliminate them by overlaps or by placing them along the edge.

Slide the bark back on the table and lightly paint the sky area of the paper with paste. Replace the large, tissue-thin pieces in their original positions. Use the glass and frame to check the effect and add clouds if necessary.

The third and easiest method is to choose a thicker piece of bark with realistic sky colors and perhaps the cloud pattern already on it.

If the whole sky is to be made with a single piece of bark, then pasting down is simple, but if there are four or five carefully integrated pieces then they should be pasted without disturbing their relative positions. Anchor one side of a piece of bark with the heel of the left hand, lift the loose end a little with the left hand finger and thumb and

THE COVER PICTURE

The trunk and nearby branches of the massive tree were cut from bark so furrowed by insects that it closely resembles the bark of our Scribbly Gums. The characteristic markings on the trunks of these trees are also made by insects.

Junctions between the smooth bark of the younger branches and that of the older wood were disguised where necessary, by almost thread-thin pieces of bark. Two bark knots were added and a hot poker used to blend them with the trunk, as well as to shade the tree and emphasise the dark hollow at the top of the trunk.

Tissue-thin bark, crumpled to suggest wind-blown dry grass forms the pasture where the cows are grazing. The old fence and sliprails are torn from brown bark and shaded by burning.

Every piece of bark, even the small flimsy ends, was pasted down securely. Because of the bulk of the large tree the picture was pressed face up by covering with soft plastic and a tray of very heavy books.

When dry, it was lightly sprayed four times with hair spray. This seals the bark without making it shine. It was then framed without glass.

dab paste under it in several places. Continue in this way until all the sky bark is pasted down. Fine clouds may be added but if the sky pattern has been chosen well, no further additions will be necessary.

Grey bark is sometimes mottled and angry and makes realistic storm skies: cream, pink, white and a mixture of these colors are good sky material. Avoid at first using the buff colors for skies, you will have plenty of those shades in the rest of the picture. Later, with more experience, pale brown tints can be used for sky with the darker buffs in the body of the picture to give a remarkably realistic effect of an outback Australian landscape.

A WORD OF WARNING: Only when the sky is pasted down and finished should the rest of the picture be assembled or the result could be frustrating confusion.

THE LAND MASS

Texture variations in a bark picture are as important as its variations in color. Flat, smooth monotones when used exclusively, can occasionally produce the effect aimed at; in sandhills and desert scenes monotones are necessary, but used in other landscapes they could appear colorless.

The type of Australian countryside usually depicted in bark pictures has distinctive changes in color and form. These contrasts can be emphasised when rougher and more spectacular pieces of bark are coupled with those of a smoother and more tranquil texture. Sometimes bark is found with undulating parallel markings. A piece of this bark forming a foreground rise against the grey of a distant mountain will clearly demonstrate the value of contrasts. When using bark marked like this with the pattern distinct and even, as it frequently is, the hill will appear to be furrowed by a plough.

Even though a piece of bark at first seems perfect for a position, don't be satisfied until several experiments have been made. Reverse it, place it upside-down or on its side, then move it to several other places in the landscape. It is likely that the original choice will be found to be the best place for it and it can then be returned to that position with confidence.

No paste should be used during this try-on period. Begin with the horizon hills and mountains, placing each hill and rise on top of the more distant one, overlapping at this stage as much as you like.

The headband of this hunter resting is of red geebung bark and black geebung is used for the hair. Burnt bark forms the body. All pictures on this page are by Zillah Needham, Orange.

Right: Burnt bark is used to form the body of the Aboriginal warrior and fine white bark for the warpaint and the shield decoration.

Left: The large white trunked eucalypt of Central Australia is frequently seen in landscapes from that area. (After a painting by Albert Namatjira).

The gentle dim-greys and green-greys give good perspective when used for far away land masses, but bushfire-blackened bark can also be used effectively in a distant skyline. Darkened bark will separate the pale tones of two adjacent barks. If the edge of the bark forming the lower hill is burned a little with a match or candle it will then show up well.

Only when all the hills and mountains are in place and a final inspection with glass and frame has proved satisfactory, should any pasting down be done. Every effort should be made to paste the bark down while it is still in its original position.

If these carefully arranged pieces are removed for pasting as the flimsy sky bark was, it would be hard to replace them exactly where they were. It is always difficult and sometimes impossible to remember their precise location and a second arrangement could be cramped and less spontaneous.

With the picture flat on the table, steady the bark with the left hand as described when pasting skies, and gently lift the loose end of each piece of bark sufficiently high to place several dabs of paste under it. Press it back into position.

If overlaps are too great the surplus can be torn off. Using the same anchoring method as for pasting down, hold the piece of bark in place and tear away superfluous underlaps. A little overlapping is always necessary to cover the paper—too much could cause the picture to be too bulky to frame.

It is inadvisable to add any embellishment to the picture until the pasting down is completed. If a tree is planned or if colored rocks or ground growth would be an improvement, place them in position now.

As contrasts in color and texture give definition, interest and impact to a picture, strong rock shapes and trees have a similar effect. If a tree is placed in position too early the tendency is to build the picture round the tree. This can be inhibiting and the picture could soon lose its vitality.

When the tree is added after the land bark is in place, it can then be moved about at will until its most effective position is found.

Inland areas of Australia have hard strong outlines and vivid colors. Many of these colors are found in paperbarks. The characteristic deep reds of the soil and the browns, fawns and blacks of the tree trunks are all available and can be used effectively to form an outback scene.

The white trunked desert gums and the wonderful rock formations of Central Australia dominate this picture. It is after a painting by the famous Aboriginal artist, the late Albert Namatjira. Made by Zillah Needham, Orange, N.S.W.

Sometimes a picture can capture the feeling and mood of the outback successfully if small rather than large pieces of bark are used. The composition must be planned beforehand and the small pieces torn as needed and placed side by side, overlapping a little, to form a hill or tree. When shading is necessary, this is done with other small pieces, darker or lighter in color placed in the right position. Trees, land features and figures of any size and shape can be built up in this way and a strong and striking effect can be achieved.

Water and sky are made in the same way and can be shaded with many exciting color combinations introduced.

The effect when this method is used is rather similar to that gained with heavy oil paint.

Seascapes are on the whole more difficult to construct than landscapes, but with care and patience many beautiful sea pictures can be made.

The sky should be planned carefully with thought given to the compatibility of sky and the movement of water. A stormy cloud-covered sky above a rough complicated sea could be overpowering. A more balanced picture could result with the most of the action either in the sky or the sea but not in both.

Seascapes usually include a few rocks or islands as contrast or a cliff to add strength. On the other hand, a strip of foreground sand with spent waves lapping over it may fit the quiet mood of the picture.

For added realism, small foreground islands and rocks need spray and foam at their base. Large waves dashing against a rugged cliff with accompanying white spray will give splendid impact to a sea picture.

White seagulls on the beach or resting on contrasting rocks show up well. They look graceful, too, in flight, silhouetted against the sky or water or sweeping low over the sand. Their clear clean lines suggest cutting rather than tearing and for the perfectionist the standing seagulls can be given red legs and beaks.

It must always be kept in mind that the horizon of the water itself will be a straight line. If this line is unpleasantly hard or if it appears to cut the picture in two it can be broken by a few islands or a boat or by some ominous clouds rising behind it. A few tall shore trees could assist in breaking this unattractive line and improving the composition of the picture.

LAKES AND POOLS

If still water is planned for the picture, place the selected piece of water bark in position after the sky has been pasted down. If possible have the color of the water (or at least its surface pattern) differing from that used in the sky. For instance, in nature a smooth grey pool or lake could be expected under the mottled grey of a stormy sky.

As there are no blues in paperbark, pale colors depicting water look more realistic than darker shades. Greys, creams, whites, yellows and

Except for isolated areas of the sky and the sea, this carefully constructed seascape was made with small pieces of bark. No other additions were used. Made by Hilda Hicks, Canowindra, N.S.W.

pinks are available so the choice is reasonably large and a thorough search for a realistic piece of water bark will be well worth while.

When the water's shape and size has been decided, paste it down; its surrounding hills and slopes and grass banks will overlap and be added later.

It is now time to search through the remaining bark and carefully examine both sides of every likely looking piece. Gather up a selection of these pieces and carry them to the work table.

Green bark makes effective grass banks near water, especially in the foreground of the picture, with rushes and reeds round the water's edge. Colored rocks can add interest and small islands will break up a too flat surface in a large expanse of water. Small boats and canoes have the same effect. If there is a darker shading towards the distant shore it can be made to appear as a reflection in the water if an appropriate land mass is placed above it.

If, after critical examination, an already pasted down landscape looks too heavy and uninteresting, try placing a piece of water bark over some of the foreground land feature. This bark's shape and color must be sufficiently pool-like to give a definite illusion of water. If this added lightness improves the composition of the picture, the bark under the water bark can be easily removed as the paste will still be damp. The water bark can then be slipped under its heavier surroundings and pressed into position. If this pale water bark is merely pasted on top of the heavier bark it could appear patch-like and be quite unrealistic.

Few rivers can be seen in nature with an uninterrupted course from infinity to the foreground of a landscape. A curve in its direction, a rise in a hill or overlapping vegetation will almost always block out part if not most of a river's course. How much more interesting the view is when this happens.

As in nature the river in a bark picture looks very much more attractive in part than in whole, it can wander across the picture at an angle and be wide or narrow as the picture demands. If the river's foreground bank is broken in places by bushes, reeds, rocks or even a tree or two, this will help to eliminate the tapelike effect. With perspective in mind, smaller vegetation scattered along the distant bank would also assist in forming a more realistic looking stream.

WATERFALLS

The introduction of a waterfall to a bark picture will always have a dramatic influence, but care should be taken to see that it is integrated into the picture, not just pasted on. A slender tree branch passing in front of the water, a few overlapping fern clumps or bushes at its sides or an outcrop of rocks will assist in making the waterfall appear as if it belongs to the landscape.

An interrupted fall of water will have graceful lines when it slips smoothly over a cliff edge and breaks its fall on a jutting rock ledge lower in the picture. Some foam or spray may be effective here before the waterfall continues on its way out of the picture, or is blocked out by the foreground vegetation. Usually a waterfall looks more realistic when its bark is pasted in place before the surrounding landscape bark is added; on the other hand if the fall is to be slim and delicate, white flimsy bark torn to shape and arranged over the already completed background will give a lacy illusion unattainable when the first method is used.

There is no hard and fast rule as to which bark should overlap, the shape of the waterfall and the type of bark used will be the guiding factors.

When the water's position is decided, paste it in place, then if necessary add environmental rocks and vegetation.

CHAPTER FIVE

Additional materials that can be used

ALTHOUGH A TRUE BARK picture is made only with bark in its natural colors, the addition of selected lichen, mosses, grasses and ferns give interest and character to what might otherwise be an uninspired landscape.

LICHEN, MOSS AND FERN: The choice here is unlimited and lichen which grows on trees and rocks during the winter months is to me the most useful and fascinating of all the materials available. The wonderful detailed beauty of a lichen's formation is so often passed by without a glance. Its shape and color varies with the climate and the locality in which it grows. There are many shades of whites, greens and browns and even bright yellow and brick red; unfortunately the last two lichens are usually the most difficult to remove from the host-rock or tree.

Lichen is especially beautiful when in flower and the shape of these flowers will remain unchanged when dry. Its color has a better keeping quality if picked early in its life cycle.

Some mosses are an attractive addition to a picture. They too, dry well and like lichen keep their color if picked in the early winter. Spagnum moss grows on damp shady banks, the color of its young plants is emerald green with the root area a rich golden brown. If its color fades a little over the years it does this gracefully and the shape will remain unchanged.

Fine native ferns and some grass seeds are effective in a bark picture when placed behind small rocks and round the roots of trees. They then assume the size and importance of large palms. Ferns should be pressed but lichens and mosses need no careful drying and storing. Most of the lichens remain pliable when dry and can be pressed flat with fingers. Occasionally a soft growing lichen will harden considerably as it dries. It can be made plastic again with a little warm water and then flattened in a book between newspaper sheets: the same method as for pressing flowers. In this case there is no need for the lichen to remain in the book until it dries. After 24 hours remove it and allow it to dry in the air; its very-flattened look will disappear and a slightly rounded and more natural contour will replace it.

Some lichen can be pieced together to form small trees or added to cut-out tree shapes to depict realistic leaf growths. On the ground they can integrate a too stark feature, such as a rock or tree root, into the surrounding scenery. As small bushes or distant scrub growth they have great value, but when a certain lichen or other material is used to represent leaves on a tree, be careful not to use that same material as ground growth. The effect will be unreal and unconvincing. With so many different materials available this repetition is unnecessary.

REEDS: Reeds behind foreground rocks or near water look effective and quite life-like. Fine grasses can be used for this purpose, but between the layers of some sheets of bark, thread-like strips are found with colors varying from pale to dark brown and even henna. Gather up a small bundle of these about two or three inches long, double them near the centre, leaving their ends uneven. This doubled end is now the root of the reed clump and if uncontrollable can be kept in place with a twist of cotton. For a less compact clump, no tie is needed.

Native grasses are occasionally found that are green in color, but with dark brown nodes. If these are cut or pulled apart with care, they can make realistic bulrushes. Dip the root ends in paste, tuck them behind a rock or under the foreshore bark and press them with a finger for a few seconds to insure contact.

CUTOUTS: Many and varied are the objects suitable to be cut out and added to bark pictures. Houses, huts, sheds, fences, birds, animals, trees and human forms all have their places. Sometimes these additions look more natural when torn rather than cut. Ancient wooden fence posts are weather-worn and realistic when torn from rough grey bark, but on the other hand small birds need the clean precision of cutting and trees, too, are usually more graceful when shaped with scissors. The actual lines of the article will always control the method to be used in shaping it.

For cutting, bark is easier to manage when it is soft and pliable and fairly smooth. It may be necessary to draw the outline of the object on the back of the bark or to cut from a pattern. Changes in bark tonings can be used for shading and full advantage should be taken of the value of color contrasts in cut-outs.

The addition of too many accessories of this nature will be overwhelming, but, used with restraint, cut-outs can be helpful in the composition of a bark picture. Two or three black crows will fit in well in an arid outback scene and few landscapes are successful without the addition of a tree or two. Human figures should be used with caution, but huts and fences rarely look out of place and are not difficult to make.

Some aboriginal women when making bark pictures asked me to cut them a few aboriginal figures. A boy holding a lizard by its tail and an old man squatting by the fire proved popular and for a few weeks they all wanted these figures in their pictures. They teased each other in a good humored sort of way, saying, "That is your ancestor".

TREES AND LEAVES: Tree shapes may be torn from soft bark or formed with tiny bits of torn bark pasted in a tree's sketched outline. A third way is to cut the trunk and branches with scissors, and a fourth is to use natural growths to represent trees. Black sedge, silver pampas plumes, ferns, even small twigs and seaweed are used for this purpose.

The torn or cut tree can be shaded by burning or with darker bark. For the pieced-together tree, tiny torn bits in contrasting colors make effective shading.

Leaves added to tree shapes bring life and vigor to a flat picture. The ways of depicting these are numerous. Green bark can be torn to represent hanging clumps of horizontal leaf groups, while bark in varying colors torn into very small particles, even minced, has considerable realism as foliage. Even autumn leaves give no problem; yellow bark can be torn into tiny bits and red geebung bark crumbles easily. These colors combined with browns, fawns, and greens are all the shades needed for an autumn forest scene.

Using this method, small particles can be touched with adhesive and placed in position with forceps or sprinkled over an area already lightly painted with adhesive and pressed gently, for good contact, before removing the surplus. Do this by holding the picture vertically over a container and tapping it. The pieces which fall may be re-used.

Lichens are numerous and have good texture and color; they may be torn to shape and used as leaf groups. Some mosses pressed into flat shapes serve the same purpose. Sponges too, have their uses. Cut them with a sharp knife into thin slices; sponge substance has a pleasantly light appearance when depicting leaf clumps.

Seaweed in its many hues can also form effective foliage. Pressed flat and dried, it can be broken into suitable shapes and pasted in place over the tree branches.

Try crumbled dried she-oak needles for light tree foliage or the seed heads of kangaroo grass while small finger sponges fit well into a Mexican desert scene.

The desire to depict leaves in new ways will bring forth new ideas and with the help of your imagination the field is wide.

SEAWEEDS: Soft rock growth in attractive colors are found along the coast of Australia and a wide variety of seaweeds mixed with the debris of high tide lie scattered on the beaches. Some seaweeds have a unique tree-like structure and others resemble ferns. These can be useful in a bark picture to represent small trees or the leaves on a large tree.

The curves that a branch of seaweed assumes when lying undisturbed on the sand are often graceful and suitable to use without change, but if a different line is needed and the seaweed is dry and hard, it can be softened with warm water and its fronds rearranged in a fold of newspaper. It can then be dried in the sun and very little weight will be needed to flatten it.

If the seaweed is large and its formation suitable, it can be broken into small bits and used to represent leaves on trees either as horizontal clumps towards the ends of the branches or falling softly as willow foliage does.

When introducing seaweed into a picture to depict leaves on trees, the variety used should not be repeated as low growing vegetation. Sponges and other seagrown flotsam may be useful alternatives. Tiny finger sponges have a cactus-like formation and algae and other sea growths are to be found in a wide range of colors and shapes. Used with restraint they can add color and interest to a picture.

These sketches of single trees and tree groups can be traced on to graph paper, then enlarged to copy directly on to the bark's surface. Large leaf masses look well when formed with torn bark, lichen, moss or cut sponge; smaller leaves from bark fragments or from minced bark.

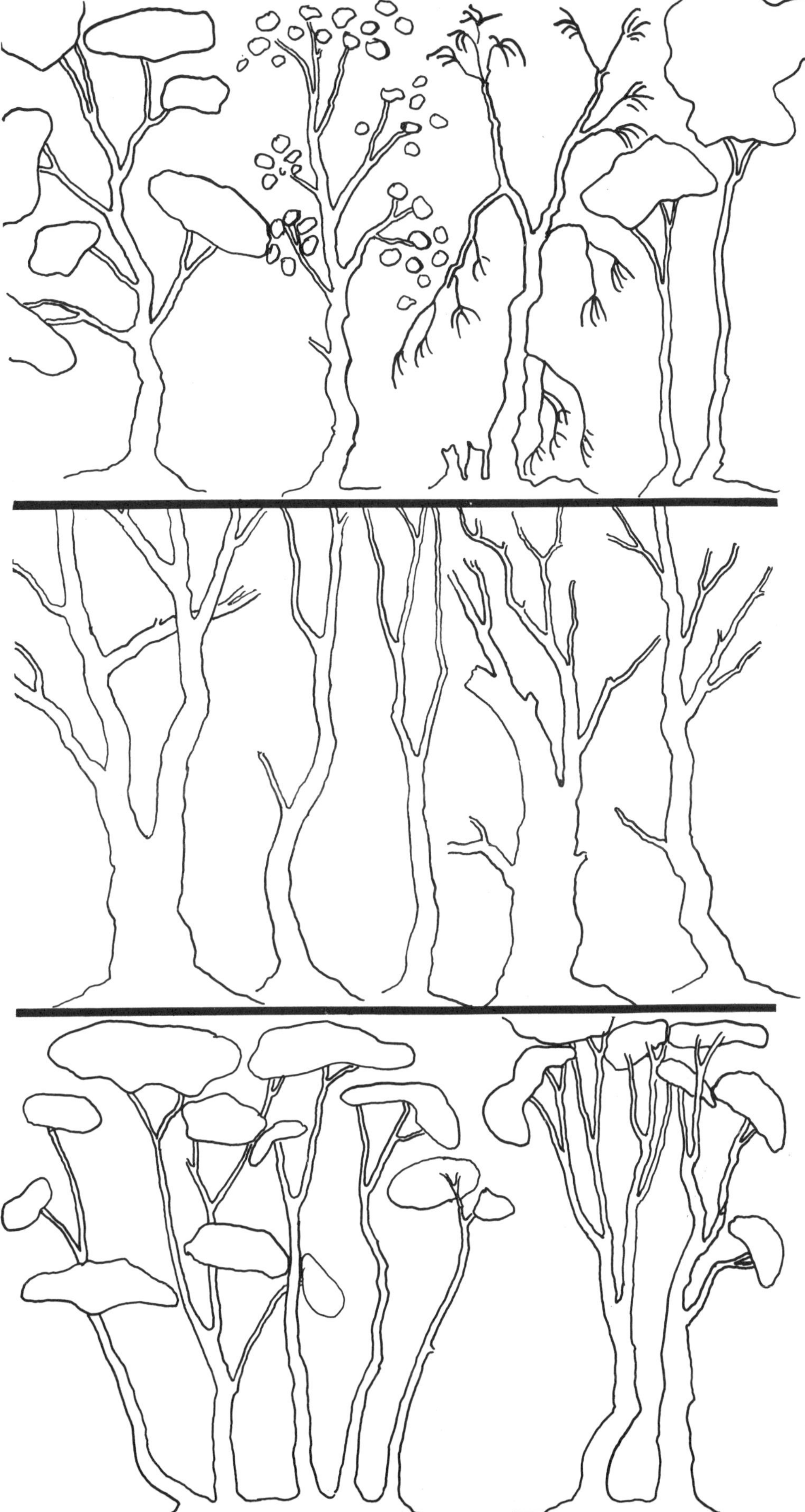

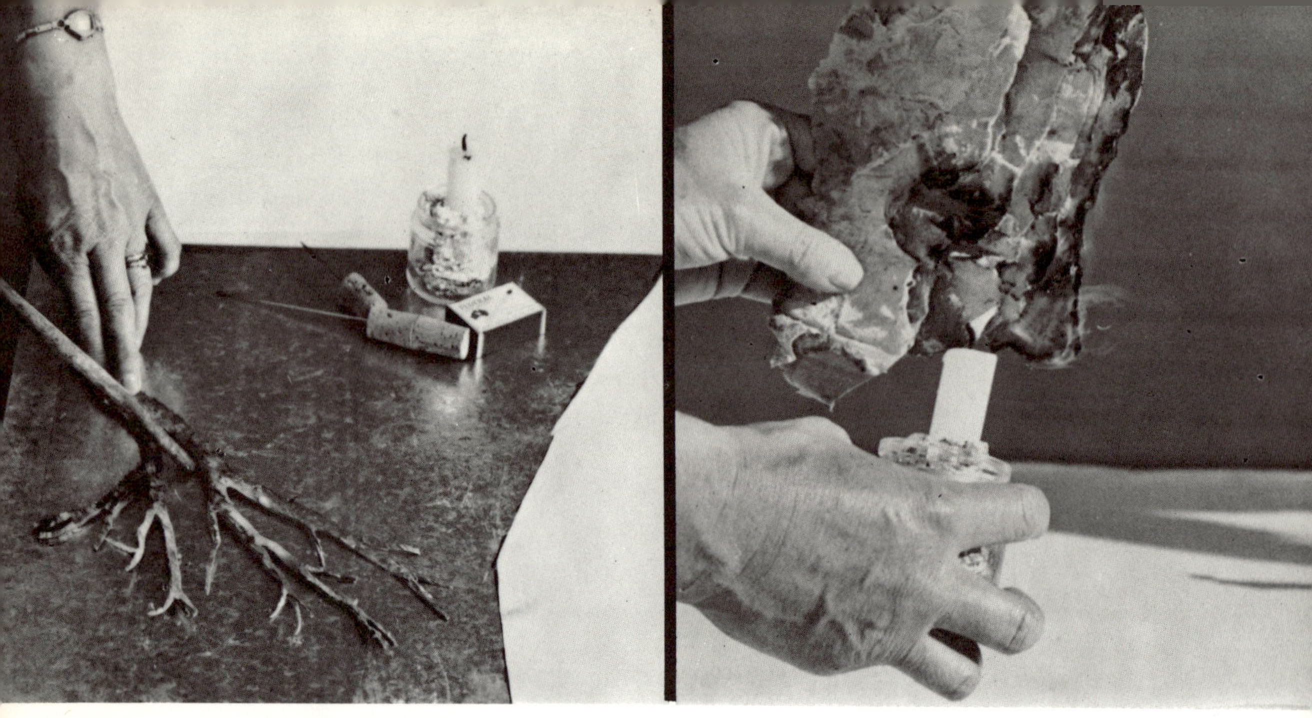

A tree can be shaded with a hot poker. Flat iron is used to protect the table.

For burning a piece of bark a candle is easier to use and safer than a lighted match.

A realistic underwater scene can be made using green colored bark for the background with seaweed trailing upwards from a sandy bed. A few flat shells, cutout bark fish and perhaps a beach-dried starfish or baby sea-horse would complete the picture.

BURNING BARK: In most collections of bark some bush-fire burned pieces will be found. These are of great value in a picture and can be used to represent such things as distant mountains, hills, rocks, trees, burnt stumps and even crows.

A bushfire seems to produce the best burnt bark, but if none is available a suitable substitute can be made by carefully burning bark pieces with a candle or poker.

Too much burnt bark can spoil a picture and if all the hills are outlined with black bark the effect will be artificial. On the other hand a little burning can improve an insipid area quite dramatically.

Loose tea tree bark is highly flammable especially in the heat of an Australian summer so care should be taken to see that the candle used is safely wedged into a metal or glass holder packed firmly with aluminium foil. The candle's flame should be extinguished when not in use.

A lighted match is quite useful for darkening a small area, but a match burns quickly and could be dropped from singed fingers into the bark pile. It is safer to use a candle and wise to have the bark-burning equipment away from the work table.

Damping the bark will help control the flame. This method could be safer and easier for beginners. I seem to harp on the safety angle when burning bark, but so often craft groups operate in places where the fire hazard is outstanding and a bark-covered table in the heat of summer could flare up in seconds.

If a piece of bark is to be completely or partially blackened, try lighting its upper end and when the flame spreads, quickly turn it upside down at various angles directing the flame to the part to be burned. The moment the bark is sufficiently burned extinguish the flame or it will devour the bark which will then be useless. I have never had my fingers burnt in the slightest degree blackening bark this way. A damp cloth can be kept handy for dowsing the flame, though blowing is usually enough.

If the edge of a hill needs darkening to distinguish it from similar shades, dampen the bark if you wish, then run the edge through the candle. Small flames can be extinguished by pinching.

Shading the trunk or branches of a cut out tree shape if done with a candle can be hazardous; should the bark burn too freely the whole tree will be reduced to ashes in seconds. A more controlled result can be obtained by outlining the tree with a very hot poker using the smooth surface of the stove or a sheet of flat iron to support the cut out tree.

When a more rounded trunk is wanted, gradual shading of these darker outlines towards the centre of the trunk will produce this effect. If the tree is to be a dead one and bushfire burnt, then the poker can be applied more liberally and all or almost all of the tree's area blackened.

A poker will make smooth black shadings, but when finer lines are needed a steel knitting needle or bodkin inserted in cork can be used with great accuracy. Effective line drawings can be made on smooth bark with these tools. Christmas cards and gift tags using this method for decoration are both novel and interesting.

CHAPTER SIX

How to finish off your picture

WHEN THE PICTURE is pasted down and no further alterations are necessary, hold it face down supported by the glass and carefully cut away all overlapping bark. It is at this stage that accuracy in cutting the background paper is appreciated. Damp backing paper together with layers of damp bark are difficult to cut with precision, whereas the bark itself can be neatly trimmed along the edge of the paper.

Now support the picture behind with its cardboard backing and add glass and frame. Place it in an eye level position facing a good light.

I think it is a sound idea to have a cup of tea at this juncture or to tidy up the bark table. You will then go back to the picture with a fresh outlook. Examine it from all angles and distances. So often obvious faults will be seen which were not noticeable at the work table. As the paste is still damp these faults can be corrected easily.

If thin layers of bark have been used, the frame should be removed and the picture placed face down on its glass or other smooth surface. Cover it with a double fold of smooth surface towel and place a flat-bottomed tray or board over this with some heavy articles on the tray. Leave it over-night. Examine the picture next day for missed faults and allow it to dry uncovered in the air. A day in hot summer could be enough to dry out the picture; a little artificial heat could speed up the process in winter.

When the picture has been constructed with bulky bark, and the tree trunks are of impressive thickness, this pressing method is unnecessary and inadvisable. Usually this type of picture will dry well in the air, but if any pressing is necessary place it on a flat surface, bark

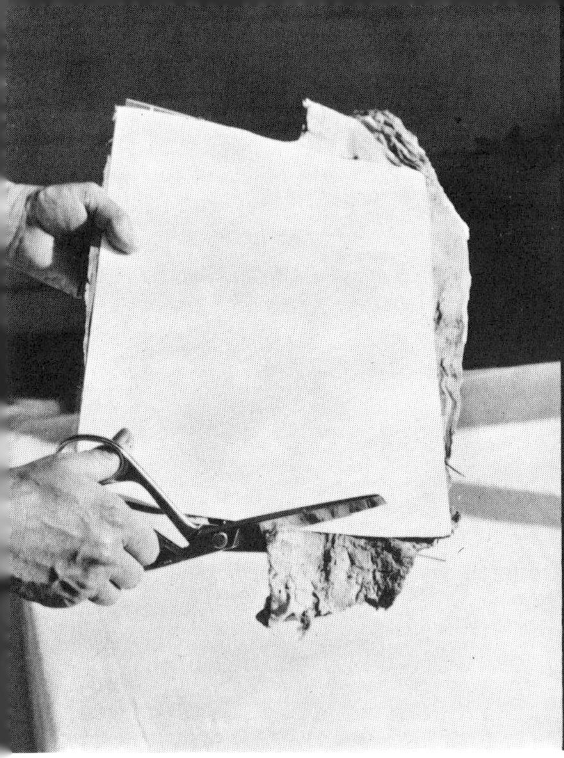

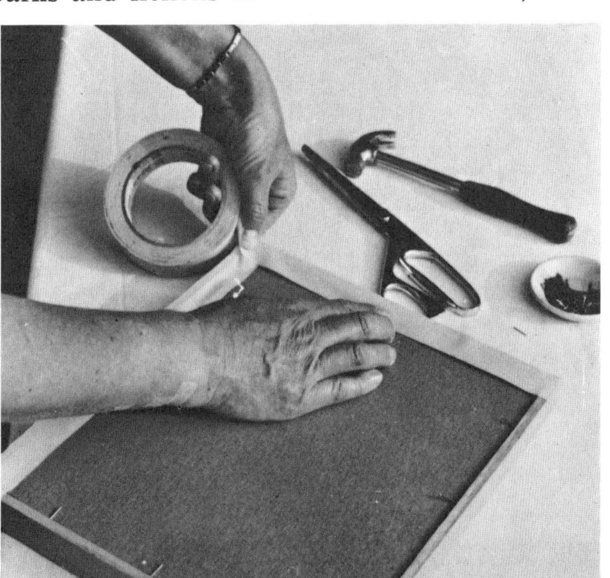

When the paste is dry, reverse the picture and trim the bark along the paper's edge . . .

. . . cover the picture with the cardboard backing and tack it securely in place . . .

side up. Cover the bark with soft plastic, then a folded towel and a weighted flat-bottomed tray as described above.

With constant handling with paste-covered fingers, the glass by now will be anything but clean. A scrub in clean soapy water and a polish with a clean dry towel will be necessary.

Do not frame the picture until the paste is quite dry. A little alum behind the picture or a spray with insecticide will discourage silverfish and other insects. If the picture is to be sent overseas, a really strong insecticide should be used. Tack the glass, picture and firm cardboard backing in the frame and seal over tacks with masking tape.

The maker's signature can be added to a small typed description of the picture pasted on its back; for example, "Made entirely with Australian paperbarks, geebung barks and lichens in their natural colors, by ...."

. . . when the picture is securely tacked in place, seal over the tacks with masking tape or brown paper.

Ideas for gifts and greeting cards

BARK PICTURES THEMSELVES make wonderful gifts, of course, but there is no limit to the many ways bark can be used to make something truly personal for a friend or relative—and be useful too.

Here are just a few ways bark can be used:

ROSES MADE FROM BARK

When choosing a flower to make from bark, the wide variation in shape between the bud and full bloom, makes the rose an attractive model. Chrysanthemums with gently curved petals, daisy shapes and the Hawaiian Wood Rose are other suggestions. The brown variations found in the wood rose are traditional bark colors; its large seedcase can be formed by using a circle of soft brown bark packed with cotton wool, its edges then gathered together and tied.

The Wood Rose's character can be changed completely when a large gum nut is used for its centre or the smooth orange seed of the native Burrawong palm. These seeds retain their vivid color for about three years.

With single daisies, thin petals tend to droop as finely tapered bark does not have the strength to support the petal's weight. A petal cut wider at its base can be folded to give this necessary strength.

For bark flowers, choose the soft pliable suede-like bark that is found in abundance in all the Paper Bark areas of Western Australia, Northern Territory and Queensland, but is less common in the southern regions of New South Wales. Pale pinkish fawns and buffs are the most likely shades, but creams and white are not uncommon.

Avoid bark with too many tissue-thin layers. When this type of bark is curled and shaped, the flimsy layers tend to separate and the petals look untidy. Should this happen, a little adhesive on a toothpick smoothed between the loose layers and a pinch with the finger and thumb, will bring them back to their original, compact state.

To form a rose, cut the petals in two different sizes. A few extra petals of each size gives a handy variation of shapes and any left-overs can be used for the next rose.

Dip the petal shapes into hot water for a few seconds and place them in a fold of damp cloth where they will stay moist while the centres of the roses are being prepared.

Bark Roses when used to decorate wall hangings need no firmly attached stems. When roses are to be arranged in a vase, tough dried rose twigs make good stems, but strong flower wire is more adaptable when curves are wanted.

Have ready the stem material of required length and some adhesive, as these will be needed early in the assembly of the roses. You will also need brown-green florist's tape, cotton or fine flower wire, a medium sized knitting needle, and if available, French Flower tools and a sand bag. The sand bag can be made in a few minutes by putting a cup of sand inside two thicknesses of nylon stocking and tying tightly.

The next step is to cut a piece of oblong bark (see diagram on page 49 to form the centre of the rose. The short length of this bark should be about half an inch longer than that of the planned bud. Dip this bark in hot water and make a half to three quarter inch fold along the top long edge. Now turn both ends of the fold towards the centre of the lower edge. Overlap these folds to form a graceful opening at the bud point. The next step is to turn the lower points of this rough triangle backwards, overlapping them behind the bud to again simulate unfolding petals. A few loose twists of cotton will hold the folds in place. The bud centre is now complete and it is time to insert the stem.

Dip the stem end in adhesive and force it well inside the bud; secure it to the end folds with several very firm twists of cotton or fine flower wire.

Put the rose centre aside and prepare to shape the dampened petals. This can be done quite satisfactorily with the fingers and a knitting needle or with fingers alone, by rolling the bark edges just as idle fingers roll a bus ticket.

If a set of French flower making tools is at hand it could be helpful, but it is not essential.

To form the saucer shape of the petal use the sand bag and marble-like tool of the French Flower Kit or cup the petal in the palm of one hand and form the planned depression with the thumb or fingers of the other hand. Then shape the outside edges by curling them round the knitting needle. A twist of cotton will draw the bark together at the bottom of the petal and improve its saucer-like shape.

When sufficient petals are formed in this way, hold the previously assembled bud in one hand and place four or five of the smaller petals round it. Keep these in place with cotton and then carefully arrange the outside petals, again using cotton or fine wire to secure them. At the same time use the cotton to shape the petal ends to form the seed case of the rose.

If a widely opened rose is planned, gently press a few of the inner petals away from the centre and fold an outer petal or two in a slightly downward position. When the bark is dry cover the seed case and the stem (if it is wire) with the brown-green florist's tape.

A light spray with clear lacquer or hair spray will enrich the color without making the roses shine and will have the added benefit of strengthening the petals.

These stemmed roses look well when arranged with real rose leaves dried in borax or glycerine or with other suitable dried material. Bark roses are attractive on wall hangings, either with their own or simulated rose foliage, or as the dominating subject in a nut, cone and seed pod arrangement.

With the loosely arranged roses, illustrated in color following page 16, the three-sided seed pods of the Golden Rain Tree (Koelreuteria) were used to form the rose leaves. When the pods were divided and the seeds removed, the shape, texture and color of the pod's sides made realistic rose leaves. Small twigs of varying sizes were glued in position to take the place of attached stems.

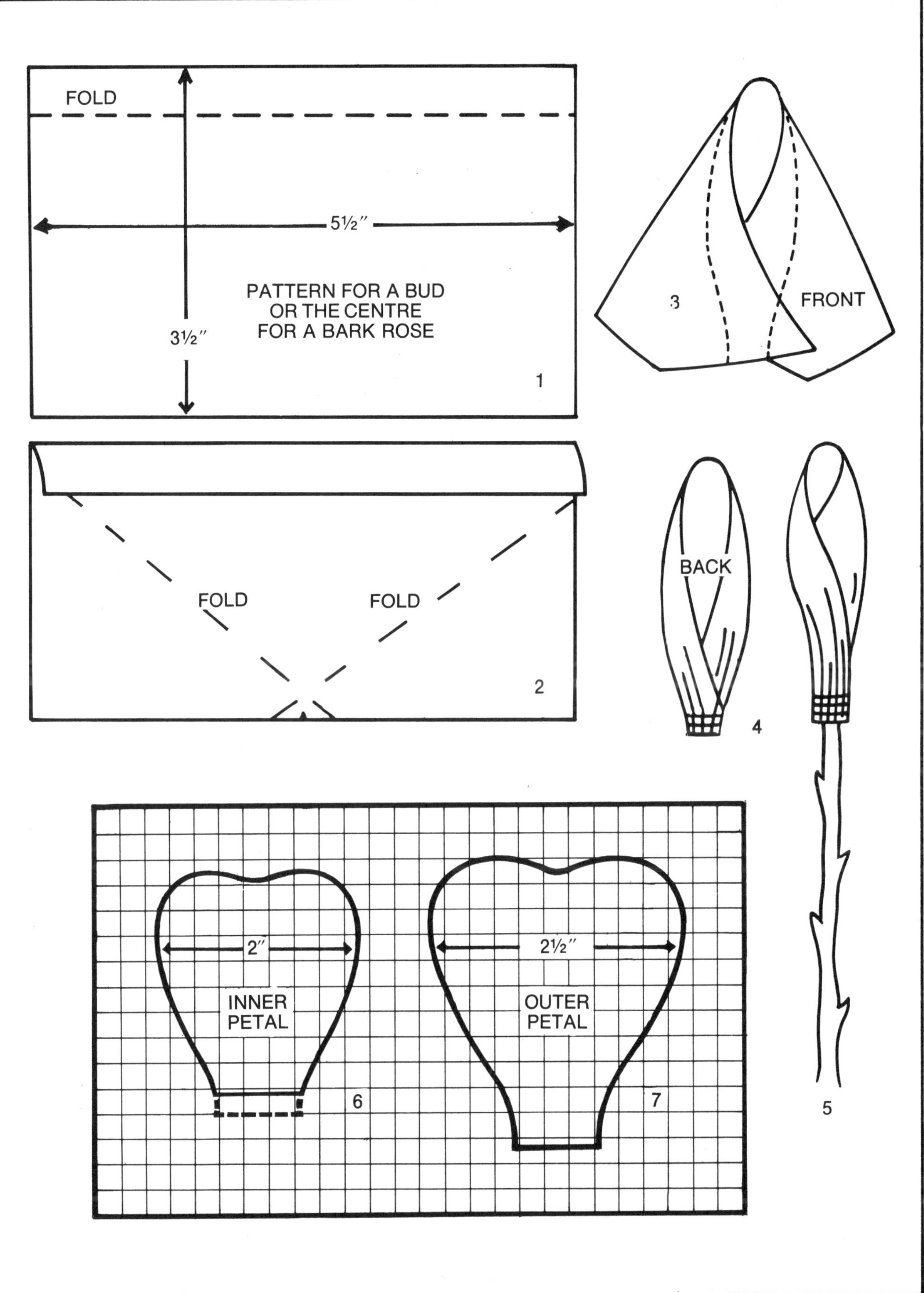

FOLD
5½"
3½"
PATTERN FOR A BUD
OR THE CENTRE
FOR A BARK ROSE
1
FOLD
FOLD
2
3
FRONT
BACK
4
INNER
PETAL
2"
6
OUTER
PETAL
2½"
7
5

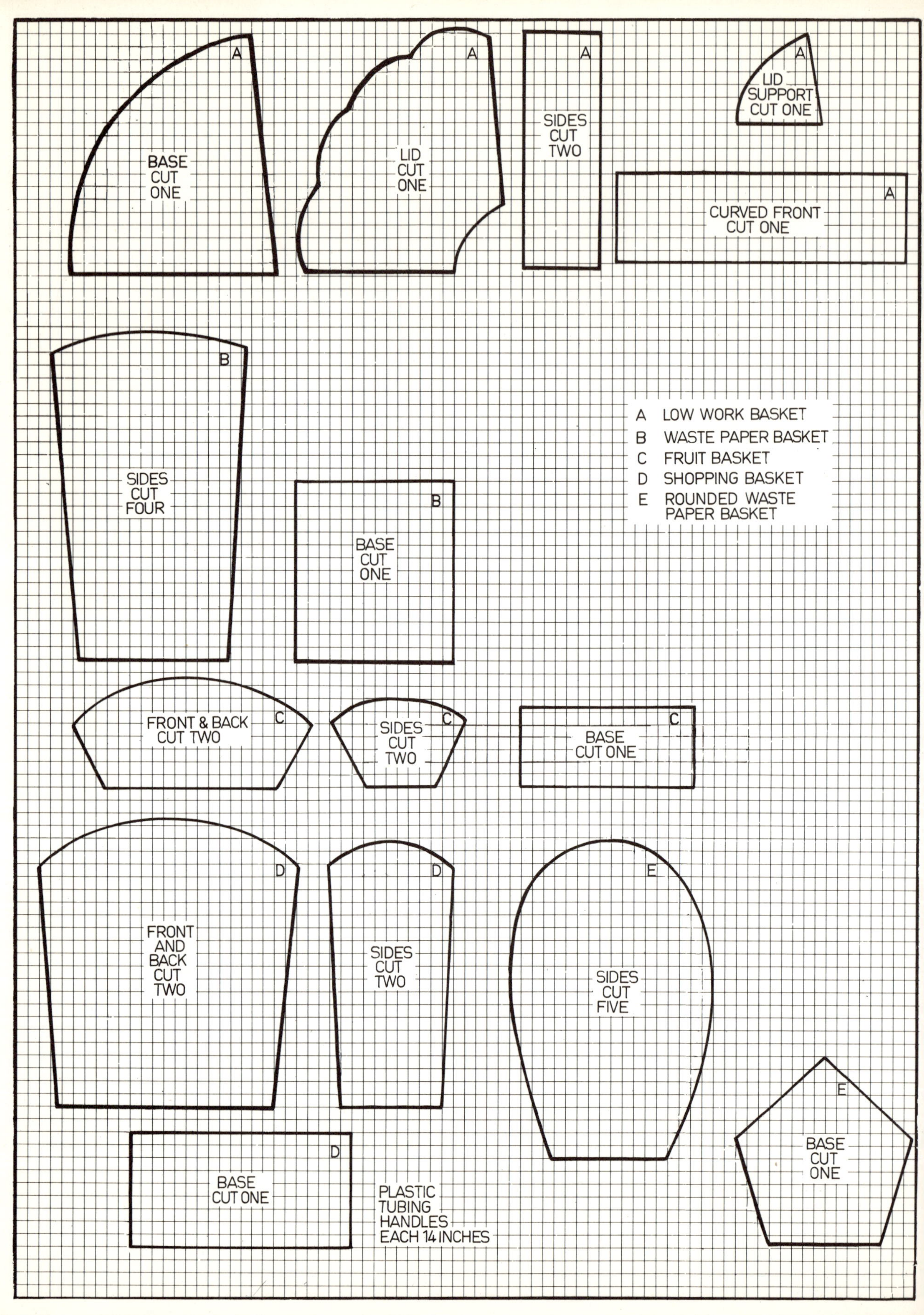

BASE CUT ONE
A
LID CUT ONE
A
SIDES CUT TWO
A
LID SUPPORT CUT ONE
A
CURVED FRONT CUT ONE
A
SIDES CUT FOUR
B
BASE CUT ONE
B
FRONT & BACK CUT TWO
C
SIDES CUT TWO
C
BASE CUT ONE
C
FRONT AND BACK CUT TWO
D
SIDES CUT TWO
D
SIDES CUT FIVE
E
BASE CUT ONE
D
PLASTIC TUBING HANDLES EACH 14 INCHES
BASE CUT ONE
E
A   LOW WORK BASKET
B   WASTE PAPER BASKET
C   FRUIT BASKET
D   SHOPPING BASKET
E   ROUNDED WASTE PAPER BASKET

WASTEPAPER, SHOPPING, FRUIT AND WORK BASKETS

For baskets made for shopping, work and wastepaper, at least four bark pictures should be made, unless the smaller ends of these baskets are to be decorated with plain bark. The backing is a strong cardboard covered with contact adhesive paper. Sheets of hard plastic will cover the picture on the outside of the baskets.

Evenly spaced holes should be punched through the hard plastic, the cardboard, the contact paper and the back. These four layers are then joined together with Italian raffia or thick crochet cotton using a glove or buttonhole stitch. Lids may be made in the same way and attached through punch holes. Handles of clear plastic tubing with a crocheted chain of matching raffia threaded through them can be secured to the basket with split curtain rings.

The pine tree on the telephone book cover was made from seaweed — the fronds finger-trimmed and pasted in place. Minced bark forms the foliage on the other trees. Use fine cord for glove stitch lacing.

The bark decoration on these small books is covered with soft plastic. Cardboard is used to stiffen the covers and contact shelving plastic or wall paper to line them. By Edwina Somerville, Tuncester, N.S.W.

ALBUM AND TELEPHONE BOOK COVERS

These covers can be constructed in a similar manner with the front, the back boards and the back strip made separately and joined with punch-holes and glove stitch.

Make a bark picture for the front cover, but the decoration of the back strip and back cover need not be so elaborate. Plain bark for the back strip and one piece of attractively colored and patterned bark may be enough for the back board, with a loose bunch of dried grasses or pressed flowers arranged across it.

Hard plastic will cover the bark areas and the backings should be of strong cardboard with plain wallpaper or contact adhesive paper attached to them. Two more sheets of hard plastic are needed. Cut these two inches shorter at their spine end than the outside sheets to form a slot or pocket to hold the book covers.

Place (1) the large sheets of hard plastic over (2) the bark pictures and back them with (3) the cardboard and (4) the contact adhesive and finally the smaller sheets of hard plastic, in that order. Punch holes evenly round the four sides of the two boards and the back strip. Glove or button-hole stitch round three sides of the boards, taking in the inside plastic pocket which is to take the book's cover. Join the boards to the back strip with similar stitching.

Paperbark can be used to decorate Christmas cards, either forming the whole picture or incorporating bark with native dried flowers. Other ways are with cut-out bark figures on a light cardboard background or as the background itself with fine line drawings.

Although these cards take a little more time than most hand-made cards, for special greetings and, perhaps, for overseas friends, they are well worth the extra trouble.

The careful selection of bark is important as the picture will be small and will rely heavily on the quality of the bark used. Sometimes as few as one or two pieces of well patterned bark are needed to achieve an attractive picture. If further decoration is wanted, a small tree, a few reeds or a clump of fine lichen may be sufficient. The card will look heavy if the decoration is over-stressed.

Using a background of plain or quietly shaded bark, small pictures containing considerable detail can be etched on the bark with a hot bodkin or knitting needle. Here the drawing rather than the background is of importance and should dominate the picture.

With similar smooth, pale bark as a background, copies of aboriginal rock drawings can be made, using a hot needle or a felt-tipped pen. One lizard or turtle or a few of the aboriginal's superbly simple fish shapes would be sufficient with the greetings written inside. Bark takes ink poorly, so felt-tipped pens are better for greetings.

Another type of card can be made using the card's paper as background. Cut a few suitable biblical figures from bark and arrange them

It is easier to cut the whole shape of the wise men from bark of the predominating color in their robes. Paste contrasting colors of folds of material, beards, face, etc. on top of this.

The bark in the flannel flower card was pasted along the lower edge only. The flowers were then arranged and their stem ends tucked under the loose bark. The flowers together with the loose upper edge of the bark were then secured in position with adhesive.
The simple shape of the vase for the buttercups was cut from patterned bark. The method used in making the card was the same as that used for the flannel flower card.

to form a Christmas scene. Color contrasts may be sufficient, but if shading is necessary fine lines can be added with a hot bodkin. When making this type of card a clear preconceived plan is always necessary. This plan may be readjusted and improved on as the work progresses.

TRAYS AND TABLES

Trays and tables can be made with a bark picture either attached directly to the woodwork or with a light cardboard backing and covered with heavy glass; beading and handles would be necessary on a tray.

When a picture is made to completely cover a small table care should be taken to trim neatly and accurately its edges as the heavy glass will be placed on to pwithout a kindly frame to cover uneven edges. Metal clips at the corners will hold the glass in place.

HARD SEED PICTURES WITH BARK BACKGROUND

Firm backing is necessary when making a picture using nuts, cones and pods as some of these articles can be quite heavy and bulky. Three-ply or hardboard are suitably strong. Cut the backing to fit the frame.

Cover the firm backing with one large piece of attractively colored bark, or several matching pieces grouped to disguise the joins. Paste this bark securely in position with a good adhesive.

Turn the backing on to a flat surface, bark side down and arrange an even weight on top. Leave it in this position for several hours to ensure good contact for the adhesive and a flat surface for the picture's background. Now place the frame over the background and hard backing and tack it in position. A hanging attachment should be put in place now as any hammering after the decoration is in place could dislodge some of the seeds.

Arrange your collection of hard seeds, dried firm leaves, cones and pods, taking advantage of contrasts in shades, colors, shapes and sizes. For heavy articles a supporting wire tie is necessary. Make marks on the bark background on either side of the heavy items while the arrangement is in position. Remove the decoration and bore small holes through the bark and three-ply. Replace the seeds and pods in their original position and wire them securely to the backing through the small holes. A wood adhesive will keep the lighter articles in place.

The background of patterned bark was pasted to a three-ply backing and the hard seeds arranged on it. The heavier seeds were wired to the backing and all the material securely pasted down. One spray of clear lacquer will bring up the colors.

When the arrangement is completed with all pieces, including those with the wire supports, securely pasted down, lightly spray the picture with clear lacquer. This will strengthen the colors without making the seeds or bark too shiny. Cover the wire ties at the back of the three-ply with a sheet of cardboard kept in place with masking tape.

FLOWER PICTURES WITH BARK BACKGROUNDS

Large smooth sheets of bark can look attractive as backgrounds for flower pictures, especially when native flowers are used.

The color of the bark is important; the paler shades are more suitable, but brown and red-brown can be a good contrast for bright yellow or the paleness of flannel flowers (see color section).

The sheet of background bark must be flat. This flatness can be achieved by damping the bark and either pressing it overnight under a heavy, even weight or by ironing it between sheets of newspaper until it is quite dry. Paste the background bark to the cardboard backing of the picture and trim overlapping edges.

Flowers take several weeks to press and dry out. Pick them as they approach maturity. They must have no dew or rain dampness present or their natural color will be impaired and the flowers could turn brown or even black. To press the flowers, arrange them between small sheets of newspaper in an old book or telephone directory and place a heavy weight on top of the book. The newspaper sheets absorb moisture and should be changed daily for four or five days, then the flowers should be left in the book for another week to dry completely.

To make the picture, arrange the pressed flowers on the flat bark surface, avoiding crossed stems and uncovered stem ends except at the base of the picture. Care should be taken to see that each flower is used to its greatest value and not obliterated by having another flower placed over it. A small spot of adhesive is necessary behind the head of each flower as the pressure of the glass may not be sufficient to keep every sprig in position. When placing the adhesive the arrangement should not be disturbed. Lift each flower head with forceps or fingers just high enough to place the dot of adhesive behind it and lower it back into position.

When this pasting is completed allow an hour or so for the adhesive to dry; wash and polish the glass and add it and the frame to the picture. Tack the cardboard backing in place and seal with masking tape.

COASTER PICTURES

Inexpensive round wooden coasters in brown and natural timber shades are available in stores. These can be adapted and used as both backing and frame for flower and seed pictures. (See color section).

The central cork covering is easily removed with a pointed knife to allow room for a more interesting bark background. Choose thin smooth plain or lightly patterned bark and carefully cut it exactly the size of the central depression in the coaster. An accurate pattern can be made by placing light paper over the recessed area and pressing around its sharp edge with a finger. Cut this circle with care and use it as a pattern to shape the bark.

Cover the depression in the coaster with wood adhesive and press the bark circle in place.

A hanging loop is necessary. Cord or ribbon may be attached to the back of the coaster with adhesive; another method is to bore

The kangaroos were cut from kangaroo skin scraps with the fur clipped to about a quarter of an inch in length. Trees cut from bark were pasted in place with rock lichen forming the foliage.

two small holes in the "frame" of the coaster and thread a leather thong through with knots to the front.

Arrange small, firm flowers or hard seeds in a pleasing group and secure them in position by lifting each flower head or seed high enough to place adhesive under it; press it firmly back in place.

Allow the adhesive several hours to dry before hanging the coasters.

BARK WALL HANGINGS FOR CHILDREN

Cut a piece of bark about 12 inches by nine inches and press it flat by damping the bark and ironing it dry between sheets of newspaper. Place this bark under a heavy book or under some weighty article and leave overnight.

Now cut a strong cardboard backing one quarter of an inch smaller on all sides than the bark backing. Join the cardboard and bark with a

Left: The hanging cord and rod are from an old calendar. The heads of yellow native everlastings and small western kurrajongs were pasted on a bark background reinforced with cardboard. Right: Accuracy in detail is important in brooches and each piece of bark or trimming must be pasted down well. Burnt bark forms the distant trees and colored coral seaweed represents foreground vegetation. The brooch is one inch in diameter. It was made by Beryl Dale, Surfers Paradise, Qld.

strong adhesive. Using a leather punch make two holes about one inch from the top and equal in distance from each side. Thread cord or a leather thong through the holes to form a hanging loop and knot it on the front of the banner.

The animals may be cut from bark and shaded, but in the two hangings illustrated the shapes of the kangaroos were drawn on the inside of a kangaroo skin and then cut out. Excess fur was trimmed off before the kangaroos were pasted in position on the bark. Trees were cut from thin bark and pasted in place and the lichens used to represent leaves were well moistened with adhesive and pressed firmly to the bark. Spagnum moss and lichen form grass clumps round the roots of the trees. A piece of contrasting colored bark across the base adds balance to the picture. A folded towel was then placed over the wall hanging with an even weight on top; this was to establish good contact with the adhesive.

After about an hour remove the weight and allow the adhesive to dry in the air.

Left: This Siamese cat was made by a twelve-year-old girl. The peachy beige of the bark used is similar to the actual color of these cats, with burnt bark forming the contrast. The cat's eyes of beads and the pearl necklace add the touch of glamor and luxury that Siamese cats demand. Made by Helen Campbell, Hervey Bay, Queensland. *Right: This letter holder and a larger magazine holder can be made in the same way.* Made by Edwina Somerville, Tuncester, N.S.W.

BARK BANNERS WITH FLOWERS AND SEEDS

Cut out the bark background choosing smooth, slightly patterned bark. Flatten it by ironing, first under a damp cloth, then between sheets of newspaper until it is quite dry. Place the bark in a book and leave it overnight to give it a long lasting flat finish.

Now cut a firm cardboard backing a little smaller than the banner and attach it to the bark with wood adhesive. Glue a hanging loop towards each corner of this cardboard and insert a stiffening rod through these loops with a hanging cord tied to the ends of the rod.

Any type of everlasting or firm flower may be used to decorate these bark banners. Small nuts, pods and cones combine well with flowers or may be used alone to make an attractive combination. The yellow ever-lasting flower heads and small western Kurrajong pods in the illus-

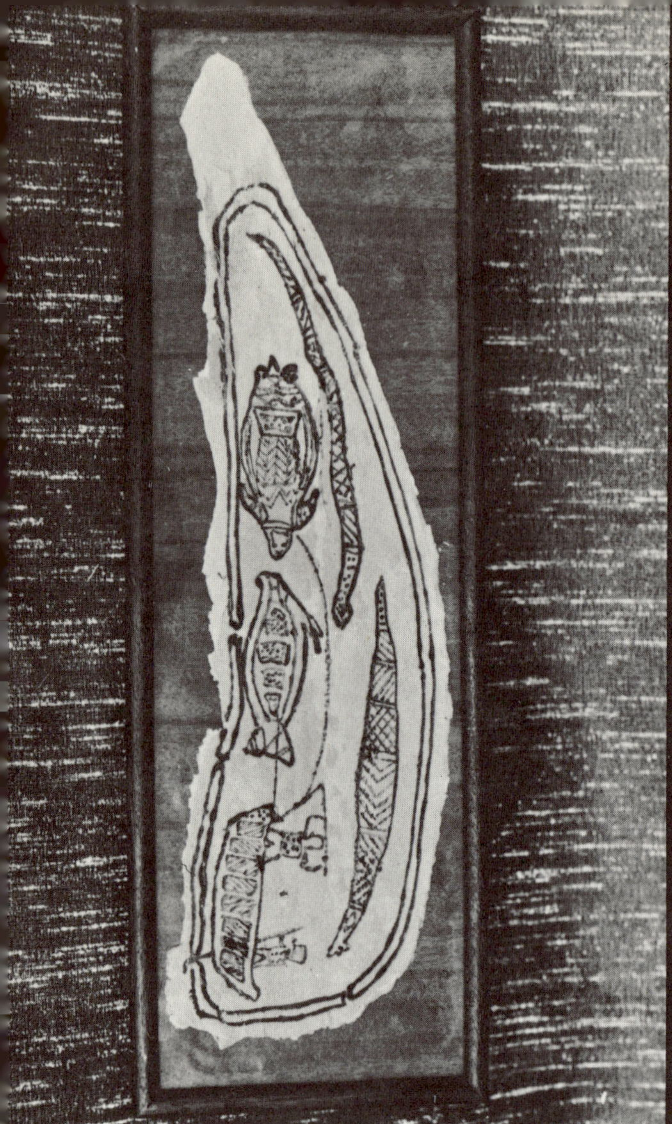

A fine felt marker is used to copy aboriginal drawings. The hard plastic cover is cut from a shirt box lid. The bark used for the copy at right of an aboriginal drawing is pasted on a background of teak stained plywood. Figures and framelike markings were burnt into the bark with a hot poker.

tration were first arranged on the bark background then made secure while still in their original position.

Do not remove a carefully assembled arrangement for the pasting down process. Lift each flower and pod just high enough to place a dab of wood adhesive under it and then press it firmly back into its place.

Allow the banner several hours to dry.

BARK BROOCHES

Metal brooch mounts are made for lapidary work in various shapes and sizes as well as in different qualities. These can be used to frame a small bark picture to form attractive and unusual brooches.

Select thin bark for its beauty in detail of pattern; color contrasts are also of great importance as so little bark is used in the construction of these tiny pictures.

Cut the sky bark to the shape of the internal oval or circle of the mount and paste it in position or use a paper backing if you wish. Arrange and paste down the land and water features using the same methods as when making a larger picture. Trees must be very small. Moss, fine grass seeds and fern-like coral seaweed in its many colors are suitable for this purpose. Tiny pieces of burnt bark can also be surprisingly realistic when used to depict distant trees.

The brooch's covering may be slightly curved and will do little to hold the bark in place. The bark and the additional trimmings should be carefully pasted down with a good adhesive as a brooch is subject to considerable movement which could dislodge lightly pasted bark.

When the decoration is completed, if a paper mount is used, trim the edges neatly and paste it securely in position on the brooch's backing. Assemble the mount, press the claws back in position and the brooch is ready to wear.

Animal Bark Pictures For Children

Flatten the bark background by damping it and then ironing it between sheets of newspaper. Attach it with a little adhesive to the frame's cardboard backing.

The animal shapes may be drawn on the bark or a paper pattern used. Cut the figure out, add eyes and shade the animal where necessary with a darker bark or by burning it with a hot bodkin.

Now place the bark animal in position on the background and add trees and other trimmings if they are considered necessary.

Allow a few hours for the adhesive to dry before placing the glass and frame over the bark picture, and tacking it in place. Use masking tape to cover the tacks and seal the picture.

Letter Racks and Magazine Stands

Although varying greatly in size, letter racks and magazine stands can be made in the same way. When wrought iron stands are used, make two bark pictures just a little larger than the iron frame and cut cardboard backings to fit these pictures.

A firm clear soft wrapping plastic is sufficiently strong to cover the bark picture. Cut this with an overlap of two inches and draw it tightly over the picture and backing, being careful to eliminate all wrinkles. Attach the overlap to the cardboard backing with sticky tape.

Contact adhesive shelving plastic is long lasting and easy to use; it is suitable for lining the bark picture. Cut this the same shape as the pictures and press it in position over the cardboard.

Using a leather punch, make evenly spaced holes round the bark pictures about a quarter of an inch from the edges for the letter stand and half an inch from the edges for the larger magazine stand.

In the letter stand treble buttonhole (three buttonhole stitches to the one hole) with Italian raffia was used to keep the edges of the bark decoration secure. The picture was attached to the iron rim with neat over-sewing through the back of the buttonhole stitching. A wrought iron magazine stand would be finished in the same way.

If the magazine stand is made of wood, contact adhesive paper is unnecessary. The picture can be finished in the same way with buttonhole stitching for trimming and attached to the flat side with wood adhesive. Another method is to paste the picture to the flat wood side and use wide masking tape to join the edges of the picture to the wood. No punching or buttonholing is needed here, but care must be taken to see that the tape makes an even line where it frames the picture. If the neutral color of the tape is considered unattractive it may be painted.

BOOKMARKS CHILDREN CAN MAKE

Bark bookmarks are quite easy to make and when children live in areas where paperbark trees grow the basic materials needed in their construction can be gathered by the children themselves. The collecting and pressing of a few native flowers for decoration is a job most children will enjoy and one that will take up very little of their time.

A simple bark landscape picture, some copies of aboriginal rock drawings or a few pressed flowers arranged on the bark are three suggestions for decorating these book marks.

Hard plastic is needed to cover the bark, but firm plastic shirt box lids are readily available and are quite suitable for this purpose.

When the decoration is to be pressed flowers or aboriginal drawings, cut the bark background and two strips of hard plastic all the same size—about eight inches by one and a half inches and shape one end to a blunt point. Choose plain bark for the drawings and attractively patterned bark for the flowers.

For drawings use a finely pointed felt pen; this will run freely and will not tear the bark. The flowing lines of the aboriginal drawings are easy for children to master.

For a small bark landscape cut a piece of pale sky bark the full size of the bookmark. Use this as a background with a few hill folds and perhaps a little moss as vegetation pasted on it. Check the back of the sky bark for appearance and if necessary cover it with a thin layer of more attractive bark. Trim the edges and cut plastic cover strips.

When the drawing, flower or picture decoration is completed place a plastic cover strip on either side of the bark and keep the various layers in place with small pieces of clear sticky tape.

With the smallest cutter on a leather punch make evenly spaced holes round the book mark about a quarter of an inch apart and one eighth of an inch from the edge.

Use heavy embroidery cotton and a darning needle or bodkin with glove stitch or treble button hole to join the layers but remove the adhesive strips as the sewing progresses.

If the sewing is started in the centre of the short straight end a tassel can include the starting thread end. Make the tassel by sewing eight or 10 long loops through the two centre holes. Bind these together with five or six twists of cotton close to the holes securing the thread end with adhesive if necessary. Cut and trim the double ends of the tassel.

COPIES OF ABORIGINAL BARK PAINTING

If a copy of an authentic aboriginal bark painting is made it should if possible be a true copy, as all the figures and motifs have some serious ancestral or tribal meaning telling a story or part of the history of the tribe.

These paintings have been reproduced frequently and are not hard to acquire.

An effective way to display a painting of this nature is to choose a strip of bark about six or eight inches wide and eighteen inches long, tapering as it would be when torn from a tree. This bark should be flawless, with beauty in both its shape and the variations in its shading.

Carefully copy the bark painting on to it with a fine felt pen or hot bodkin but do not smear or mark the background bark in any way. Dull clay-like colors should be added if the picture has these colors.

Paste the bark on a neutral-toned paper or a rough linen or burlap background; add glass and frame in the usual manner.

Paper bark trees (Melaleucas) grow in most Coastal areas of Australia but are more prolific in the upper half of our Continent.

They are found as far South as Tasmania, in the North West and more sparsely in the North Eastern quarter of that State and scattered around the Victorian coast and in parts of South Australia. They are fairly widespread in the East of New South Wales and Southern West Australia, while the Northern Territory has large tracts of Paper bark forests. Queensland and Western Australia tropical and sub-tropical zones have fairly dense growths thinning in places to scattered trees.

WHERE TO BUY BARK AND HELP
A CHARITY

Here are the names and addresses of people who will supply a well packed carton of paper bark for $4, plus rail freight within their State, also the names of the charities which will benefit from these sales.

Mrs. John Grice,
    MATCHAM, N.S.W. 2251   Crippled Children of N.S.W.

Mrs. Joan Bailey,
    "Alfriston Orchard"   Dooralong Aboriginal Scholarship
    R.M.B. 428 Jilliby Road,
    WYONG, N.S.W. 2259

The Hon. Secretary,   Country Women's Association
    Country Women's Association,
    Grafton Street,
    COFFS HARBOUR, N.S.W. 2450

Mrs. P. Hardy,   Esperance Hospital Amenities
    27 Gull Street,
    ESPERANCE, W.A. 6450

It is regretted that names of suppliers from other States were unobtainable.

Designed, typeset and composed in Australia

Printed by the Everbest Printing Co. Ltd., Hong Kong